Essential Oil Recipes for Dogs

100 Simple and Easy to Follow Essential Oil Recipes for Dogs

Julie Summers

Copyright © 2016 Julie Summers

All rights reserved.

This document is geared towards providing exact and reliable information in regards to the topic and issue covered. The publication is sold with the idea that the publisher is not required to render accounting, officially permitted, or otherwise, qualified services. If advice is necessary, legal or professional, a practiced individual in the profession should be ordered.

In no way is it legal to reproduce, duplicate, or transmit any part of this document in either electronic means or in printed format. Recording of this publication is strictly prohibited and any storage of this document is not allowed unless with written permission from the publisher. All rights reserved.

The information provided herein is stated to be truthful and consistent, in that any liability, in terms of inattention or otherwise, by any usage or abuse of any policies, processes, or directions contained within is the solitary and utter responsibility of the recipient reader. Under no circumstances will any legal responsibility or blame be held against the publisher for any reparation, damages, or monetary loss due to the information herein, either directly or indirectly.

The trademarks that are used are without any consent, and the publication of the trademark is without permission or backing by the trademark owner. All trademarks and brands within this book are for clarifying purposes only and are the owned by the owners themselves, not affiliated with this document.

ISBN:153712045X
ISBN-13:9781537120454

CONTENTS

Introduction   i

1 Recipes for Odor Control   1

2 Recipes for Itch Relief & Wound Care

3 Recipes to Repel Fleas, Ticks and Insects

4 Recipes for Calming and Relaxation

5 Recipes for Joint & Muscle Health

6 Recipes for Ear, Nasal & Digestive Health

7 Recipes to Boost Energy & Immune System

8 Recipes to improve Nail & Fur Health

Conclusion

INTRODUCTION

This book contains proven ways that essential oils are safe and effective to use in combatting common problems associated with dogs, including bad odor, dry and itchy skin, fleas, anxiety, joint pains and reduced energy. In many cases, essential oils even exhibit better therapeutic effects than other types of treatment. In fact, several studies have shown that dogs treated with essential oils gained better and faster results than those treated with alternative medicine. So if you think essential oils and aromatherapy work only for humans, think again.

Most essential oils have been found to provide a variety of therapeutic benefits—among which are anti-infectious (antibacterial, antifungal, antiviral), anxiolytic (anti-anxiety), anti-inflammatory, sedative, expectorant and immuno-stimulant. And not only humans can reap these benefits, dogs and other animals too! While some oils are believed to be toxic to dogs (such as Tea Tree Oil), there are tons of essential oils that you can use to keep your dog looking fresh, smelling good, free of fleas, healthy and energetic. Some of the essential oils that can be used safely and effectively on dogs are Cardamon, Chamomile (German & Roman), Coconut, Eucalyptus, Frankincense, Geranium, Ginger, Helichrysum, Lavender, Niaouli, Peppermint, Sweet Marjoram, Sweet Orange and Vetiver. And rest assure, all of the oil blends and recipes presented in this book include safe ingredients that aren't going to cause any harmful effects to your furry pets. After all, dogs are still a man's (and woman's) bestfriend and we surely want only the best for them.

So if you want to try using essential oils on your dogs but unsure as to which ones work best (there are so many essential oils in the market!), use this book as your guide in preparing safe and natural oil blend recipes that your dogs will also love to have.

Thanks again for downloading this book, I hope you enjoy it!

RECIPES FOR ODOR CONTROL

1 — Refreshing Shampoo

Most dog shampoos that do not contain harmful chemicals and perfumes will clean your dog's fur, but do little to nothing about their natural dog odor. This simple recipe will help remove your dog's unpleasant odors and keep him smelling clean and fresh all day.

Ingredients:

240ml all-natural dog shampoo

3 drops Sweet Marjoram

4 drops Geranium Oil

4 drops Roman Chamomile

7 drops Lavender Oil

Mix all ingredients and use the shampoo on wet fur. You will want to massage the shampoo into your dog's wet fur and let it sit for 3 – 5 minutes for the oils to do its work; make sure to cover as much of your pet's body as you can, but avoid your dog's eyes, nose, mouth and ears. Let Rinse and towel dry.

2 — Bad Odor Spray

If you need something that will get rid of bad doggy smells rather quickly then this recipe is just what you need. This recipe is a topical mixture that you can spray once a day everyday for 2 weeks. After which you will want to provide your dog a short break of a few days as dogs are highly sensitive to essential oils compared to humans. If you wish to use this spray frequently you can use hydrosol instead of essential oils.

Ingredients:

1 cup distilled water

3 drops Eucalyptus Oil

6 drops Peppermint Oil

6 drops Sweet Orange Oil

10 drops Lavender Oil

Combine all ingredients in a spray bottle and mix well. Spray the mixture on your dog's body, but avoid contact with his face, especially his eyes.

3 — Soothing Dog Shampoo

Make this dog shampoo from scratch at home and be reassured your dog is getting the best quality ingredients every time you wash him. This homemade dog shampoo is safe for dogs and will leave a lasting fresh smell. This recipe is also gentle on dog fur with lavender, aloe vera and coconut oil to condition and enrich your dog's skin and fur at the same time.

Ingredients:

4-5 oz. distilled/purified water

4-5 oz. Aloe Vera

5 drops Lavender Oil

2 tbsp Coconut Oil

2 tbsp Castile Soap

2 drops Roman Chamomile

2 drops Thieves Oil

2 drops Citronella Oil

2 drops Purification

1 drop Cedarwood Oil

Mix all ingredients and store in a pump bottle. Massage it into your dog's coat to make him feel refreshed and smelling good. Use this shampoo like any regular shampoo you can find at the pet store.

4 — Grooming Shampoo

Use this shampoo recipe on your dog to give him a refreshing look and a lovely smell. It is best used while brushing your dog, as a detangle spray or as a finishing spray.

Ingredients:

10 oz. distilled/purified water

2 oz. Aloe Vera

1 tbsp Castile Soap

2 drops Rosemary Oil

2 drops Lavender Oil

2 drops Eucalyptus Oil

2 drops Peppermint Oil

Combine all ingredients in a spray bottle and shake well. Use this shampoo to rinse your pet and make sure to avoid contact with his eyes. It is best to only spray the body while using this recipe. This mixture can be stored for up to a month if stored in a cool dry place out of direct sunlight. There is no need to use a dark colored glass for storage as once mixed with mineral/distilled water the oils within this recipe will have a considerably shorter shelf-life.

5 — Doggy Deodorizer

This has to be the easiest recipe to managing your dog's odor. This quick and simple mixture will eliminate stinky dog smell and help deodorize your home. It's best used as a quick solution to manage the odor problems during rainy days or in humid weather. You can also spray this mixture onto upholstery, however you will want to also keep the room well ventilated if used extensively around the house. The smell of bergamot oil may irritate your dog in large quantities. Make sure your dog is capable of escaping the room the spray is used in whenever they feel like it.

Ingredients:

8 oz. purified water

10-15 drops Bergamot Oil

Combine water and oil in a spray bottle and spritz it around your dog 2-3 times per week. Avoid spraying him in the face.

6 — Easy Dry Dog Shampoo Recipe

Some dogs just don't like bath time. They'll run around the house all day trying to avoid the dreaded bath. This recipe is for those special pups that can't stand getting wet. A dry shampoo is a great alternative for occasional cleaning as it gets dirt and odor out of the fur.

Ingredients:

3-6 drops Lavender or Lemon Oil

1 cup cornstarch

1 cup baking soda

Apply the mixture on your pet and massage into his skin using your hands. You will want to let the powder mixture to sit for 5 – 10 minutes. This will allow the baking powder to deodorize your dog by absorbing the dirt and oils from the fur. Then you simply brush your dog down thoroughly to get the excess dry shampoo out of your dog's fur. This is a good dog shampoo to use from time to time. However, do not use this method too often as baking soda tends to build up in the fur.

7 - Coat Refresher

This spray is wonderful for when your dog doesn't need a bath but could use a little sprucing up in the coat department. Shake the bottle well and then spray lightly all over the coat, avoiding the face. Jojoba oil promotes a shiny coat, and basil oil stimulates circulation in the skin, meaning this is not only beautifying but keeps fur healthy too. Don't be concerned about alcohol drying the fur. It's only used here to bind the oils and water, and in such a small amount it will evaporate very quickly.

Ingredients:

240 mls. Water

15 drops ethanol (vodka works for this, or perfumer's alcohol)

5 drops Jojoba Oil

5 drops Rosemary Oil

5 drops Basil Oil

Combine ingredients in a spray bottle and shake well. This spray will keep in the refrigerator for a couple of weeks.

8 - Doggy Cologne

Maybe your pup has a fancy date, or maybe they just need a little scent spruce up. This is a wonderful spray for the day before Bath Day, when a little pick-me-up is just the thing to hold you and your pet over. The scents of Pine is clean and woodsy smelling, which is especially perfect in the fall and winter months.

Ingredients:

240 mls Water

1-2 drops Pine Oil

1-2 drops Bergamot Oil

Put into a spray bottle and shake very well before application. This solution will keep in the refrigerator for a couple of weeks.

9 - Doggy cologne 2

This is another pick-me-up scent for your pup. It's got a lighter, fresher smell than the other recipe, which makes it ideal for spring and summertime. This should be applied all over the coat, avoiding the eyes and head. It will perk up the coat and add a lovely, light scent to your dog's fur.

Ingredients:

240 mls Water

4 drops Clary Sage Oil

4 drops Jojoba OIl

4 drops Lemon Oil

Mix in a spray bottle and store in the refrigerator. This mixture will keep for about two weeks.

10 - Furniture Deodorizer

Instead of spraying your furniture with chemicals to get rid of doggy smells, try this all natural furniture spray. Vinegar and baking soda make odors disappear, while essential oils disinfect and supply a much more pleasant aroma. Make sure to do a spot test on a hidden area before using all over your furniture.

Ingredients:

140 mls Water

100 mls White Vinegar

2 tbsp Baking Soda

30 mls ethanol (vodka or perfumer's alcohol)

1 drop Lavender Oil

1 drop Lemon Oil

Mix in a spray bottle and store in the refrigerator. This mixture will keep for about two weeks.

RECIPES FOR ITCH RELIEF & WOUND CARE

11 — Skin Problem Relief

This recipe will relieve itching and leaves a soothing effect. Especially good if your dog is suffering from hotspot rashes or any other form of itching related to dry flaky skin.

Ingredients:

240ml all-natural dog shampoo

5 drops Geranium Oil

6 drops Carrot Seed Oil

6 drops German Chamomile

7 drops Lavender Oil

Combine all ingredients and massage onto your dog's body during bath time. Alternatively, you can use 120ml base oil (Jojoba Oil, Sweet Almond Oil, Coconut Oil, etc.) in place of the all-natural dog shampoo. Apply the oil blend on affected areas to relieve itching. Be sure to not apply too much of the oil based version as your dog's fur will trap the oils and collect dirt.

12 — Skin & Coat Conditioner

Apply this oil blend on your dog's skin and coat to relieve dryness and itching. It also helps with dry paws, calloused elbows and under belly skin. The added vitamin E will help your dog's skin to stay healthy and supple.

Ingredients:

2 tbsp Fractionated Coconut Oil

3 drops liquid Vitamin E

3 drops Frankincense Oil

3 drops Roman Chamomile

5 drops Lavender Oil

Mix all ingredients in a bowl and store in a glass bottle. Apply on affected skin areas as needed once or twice a day

13 — Skin Soother

This simple blend helps relieve dryness and itchiness. The lavender oil will calm your dog making them less agitated about their itch. Fractionated Coconut oil is a liquid form of coconut oil, which has had a certain fatty acid extracted out of it's pure form. This extraction makes the coconut oil easily absorbed by your dog's skin and blends well with other oils.

Ingredients:

2 tbsp Fractionated Coconut Oil

10 drops Lavender Oil

Mix the oils and massage into your dog's fur. This oil blend also helps fight bacteria and calm your pet's nerves.

14 — Oil Blend Against Yeast Infection

If your dog is still scratching, he might be suffering from a yeast infection. This oil blend is effective in discouraging the growth of yeast and helps keep your dog free from further infections. The lemon oil helps regulate yeast production by helping your dog's fur and skin stay dry.

Ingredients:

8 oz. Virgin Coconut Oil

10 drops Lavender Oil

2 drops Lemon Oil

Combine the oils in a clean glass bottle and shake well to mix. Massage the oil blend into your pet's skin at least once per week to improve skin health. This mixture will last a good month or two if you place the mixture in a dark glass bottle out of direct sunlight.

15 — Oil Blend Against Skin Disease

Some dogs suffer from dermatitis, which causes itchiness and hair loss. Here's a recipe to help fight this disease. However, please seek your vet's advice before using this mixture as the cause of dermatitis can be from various reasons.

Ingredients:

120ml base oil (Sweet Almond Oil and Coconut Oil)

10 drops Lavender Oil

10 drops Bitter Orange

5 drops Oregano Oil

5 drops Marjoram

5 drops Peppermint Oil

5 drops Helichrysum

Combine the oils and mix well. Put 2-4 drops onto the palms of your hands and massage your dog's neck, chest and back twice a day.

16 — Wound Care

This oil blend helps treat minor cuts, bruises, insect bites, scrapes and other minor wounds in dogs. This blend is a natural antiseptic that keeps small open wounds clean, helping to avoid infection.

Ingredients:

120ml base oil (Jojoba Oil, Sweet Almond Oil, Olive Oil, etc.)

4 drops Helichrysum Oil

5 drops Niaouli Oil

5 drops Sweet Marjoram

10 drops Lavender Oil

Store the oil blend in a glass bottle and include it in your first aid kit for dogs. Apply the oils directly on the wound as needed.

17 - Disinfectant Spray for Healing Skin

When the skin is healing from widespread rashes, such as those caused by poison ivy or mange, there is often a time of irritation between the main affliction healing and the skin returning to normal. Scratching, biting, or licking can make healing difficult during this time, so it is useful to speed the healing process. The best way to do this is by creating a clean, germ-free environment for healing to take place. This spray will disinfect gently, while spraying it means the sensitive skin won't need to be rubbed, as that may cause discomfort.

Ingredients:

240 mls Water

5 drops Eucalyptus Oil

5 drops Lemongrass Oil

2 drops Cinnamon Oil

Shake well before spraying on the infected area. Use sparingly, for no longer than a week or two.

18 - Dry Skin

Just as people have a skin type (dry, oily, normal) so can dogs! If you notice your dog's skin looking dry in patches underneath the fur, or flaky skin especially on the least furry bits of your dog, this can help. Coconut oil is very moisturizing, and Jasmine Oil increases the skin's ability to hold on to that moisture. Massage this balm into any dry patches a couple of times a week, and it should clear right up.

Ingredients:

30 mls Extra Virgin Coconut Oil

2-3 drops Jasmine Oil

2-3 drops Rosehip Oil

2-3 drops Argan Oil

Blend together well and keep in a small jar. Because of the coconut oil base, this will remain solid at room temperature which makes it easy to apply like a balm or salve.

19 - Rash Oil

One of the most common reasons for rashes in dogs (and in people!) is contact dermatitis. This could be from contact with anything: new detergents, cleaning chemicals, or just friction. It is usually not a big deal and resolves itself, but this balm can make the process more pleasant. If the rash persists for too long though, it might be something more serious so take your dog in.

Ingredients:

30 mls Sweet Almond Oil

3 drops Roman Chamomile Oil

2 drops Neroli Oil

Pat lightly into the affected area once a day for up to a week.

20 - Promote Proper Healing of Scar Tissue

It's a sad fact of life, but sometimes dogs get hurt badly or need surgery for a myriad of reasons. A talented veterinarian will be able to minimize scarring with proper stitching technique, but you can help with that too. This isn't just a cosmetic issue; the larger and more obtrusive the scar, the more likely it is to suffer from irritation and inflammation long after it's healed.

Ingredients:

30 mls Sweet Almond Oil

1 drop Bergamot Oil

1 drop German Chamomile Oil

1 drop Helichrysum Oil

1 drop Rose Oil

1 drop Patchouli Oil

10 drops of Vitamin E Oil (optional)

Combine all ingredients thoroughly and massage into healing scars to encourage smooth and less noticeable scarring.

21 - Burn Cream

For severe burns, obviously your dog should be seen by a vet. If the burn is very mild, or if a more serious burn is in its last stages of healing, this cream can help encourage healing. Zinc and Lavender keep the area germ-free, while Aloe Juice is healing to burns. The Olive Oil will keep the area flexible and moisturized so that the skin doesn't harden during healing.

Ingredients:

30 mls Olive Oil

10 mls Melted Beeswax

10 mls Aloe Juice

1 gram Zinc Oxide

2 drops Lavender Oil

Mix thoroughly and apply once a day until the burn is healed, or for one week, whichever is soonest.

22 - Sunburn

This affliction especially affects short-haired and light-colored dogs. Boxers and their cousins are particularly in danger of sunburn. This spray will not only speed the healing process with Aloe and Chamomile, but Neroli and Rose promote healthy skin as it heals. Because it is a spray, you won't need to touch the tender area with your hands, which your pup will definitely appreciate.

Ingredients:

30 mls Aloe Juice

1-2 drops Chamomile Oil

1-2 drops Neroli Oil

1-2 drops Rose Oil

Put all ingredients in a spray bottle, shake well, and store in the refrigerator.

23 - Healing Ointment for Paws

The skin of the paws is very different to the skin underneath your pup's fur. When it is injured there can be special challenges to ensuring that healing takes place properly. This ointment will provide lots of moisturization and anti-inflammation properties to keep the ideal healing conditions present.

Ingredients:

30 mls Extra Virgin Coconut OIl

2 drops Rose Hip Oil

1-2 drops Rose Oil

1 drop Helichrysum Oil

Massage into paws as they heal from little abrasions, small cuts, or scratches.

24 - Healing Ointment for Skin Issues

As skin issues like rashes and irritation heal, the skin can remain mildly inflamed. This mixture will disinfect and reduce inflammation. Only use this for a few days at a time, as the zinc can increase irritation if used for too long. It's important to find the balance of disinfection and soothing!

Ingredients:

120 mls Water

30 mls ethanol (vodka or perfumer's alcohol)

1 gram zinc oxide

2 drops Rosemary Oil

2 drops Rosehip Oil

Mix in a spray bottle and store in the refrigerator. This mixture will keep for about two weeks.

25 - Poison Ivy

Severe cases of poison ivy should definitely be treated by a veterinarian. If your dog only has a small rash, or is on the way to healing, this ointment can reduce itching and speed the process. Chamomile is a powerful anti-inflammatory, so it takes away the itch while reducing swelling. When swelling goes down, the body's natural healing processes are able to work.

Ingredients:

30 mls Jojoba Oil or Sweet Almond Oil

2 drops Elemi Oil

2 drops Chamomile Oil

Mix thoroughly and apply to affected areas once per day for up to a week.

26 - Bruises

Arnica is a powerful oil that is known the world over for its ability to help bruising heal. In addition to arnica, Hyssop and Parsley encourage blood flow, Geranium takes inflammation down, and the act of massaging this balm in even helps by breaking up the oxidized blood that makes up a bruise.

Ingredients:

30 mls Carrier Oil such as Jojoba or Coconut Oil

3 drops Arnica Oil

2 drops Hyssop Oil

1 drop Parsley Oil

1 drop Geranium Oil

Mix all ingredients together and gently massage into affected area.

27 - Healing Dry, Cracked Skin

Sometimes the skin can become so dry it takes on a weathered, cracked appearance. Usually this is due to long exposure to water, especially salt water. If this affects your dog after a trip to the beach, or after spending time in an unusually arid climate, massage this mixture directly into the skin for several days in a row.

Ingredients:

30 mls Coconut Oil

3 drops Rose Hip Oil

3 drops Argan Oil

2 drops Neroli Oil

Mix ingredients together well and massage in to dry skin.

28 - Sensitive Skin Inflamed

For dogs with already sensitive skin, inflammation becomes a special problem. Where some of the calming oils that are normally used might help a dog with normal skin, they can prove irritating to the sensitive pet. This spray contains anti inflammatories in a simple, pure water base that is safe for sensitive skin.

Ingredients:

240 mls Water

5 drops Turmeric Oil

5 drops Frankincense Oil

Mix ingredients in a spray bottle and spray onto affected area, parting the hair if necessary. This mixture will keep in the refrigerator for about two weeks.

RECIPES TO REPEL FLEAS, TICKS & INSECTS

29 — Flea Repellant

Use this oil blend to remove fleas, as well as a repellent when your dog goes outside. This blend is best used before you discover the presence of fleas on your dog.

Ingredients:

120ml base oil (Jojoba Oil, Sweet Almond Oil, Olive Oil, etc.)

6-7 drops Peppermint Oil

4-5 drops Clary Sage

2-4 drops Lemon Oil

4 drops Citronella Oil

Mix all the oils together and drip a few drops to your dog's neck, legs, chest, back and tail. You can also apply a few drops to his bandanna or fabric based collar.

30 — Flea Spray

This blend can be used as a spray and is more versatile in use. You can happily use this blend more regularly than the blend above. The apple cider vinegar can also help with yeast problems.

Ingredients:

1 liter distilled/purified water

1 cup apple cider vinegar

2-3 drops Cedarwood Oil

2-3 drops Lavender Oil

Combine the apple cider vinegar with the essential oils and mix well. Put the mixture in a spray bottle and spray it onto your dog's body (avoid spraying near his face). You may also spray it on his bedding to keep fleas off.

31 — Flea Spray

This is another oil blend to spray on your dog to get rid of fleas naturally.

Ingredients:

8 oz. distilled/purified water

10 drops Lavender Oil

5 drops Cedarwood Oil

5 drops Eucalyptus Oil

5 drops Citronella Oil

Mix all ingredients in a spray bottle and shake well before using. Spritz on your pet regularly and store in a dark area to preserve the efficacy of the essential oils. Sunlight will damage the essential oil properties.

32 — Fleas Around the Tail

Some owners will notice that fleas tend to favor certain places on your dog. One of these places is none other than your dog's bottom. Fleas tend to congregate at the base of a dog's tail or anus area due to the smell. Use this oil blend if your dog's tail seems to be the fleas' favorite dwelling. This blend unlike the others is suitable for this sensitive area.

Ingredients:

1 tbsp Olive Oil

2 drops Lavender or Cedar Oil

Mix oils and apply it at the base of the tail to prevent fleas from harming your pet, but avoid applying it directing onto the anus area.

33 — Flea & Tick Repellant Spray

This recipe will help discourage fleas and ticks from latching onto your dog.

Ingredients:

2 tbsp Fractionated Coconut Oil

2 drops Lemongrass Oil

2 drops Cedarwood Oil

2 drops Thieves Oil

2 drops Citronella Oil

5 drops Lavender Oil

Combine all ingredients in a spray bottle and spray on your dog's coat to get rid of fleas and ticks. Avoid contact with eyes and face when spraying.

34 — Flea & Tick Collar

Apply this oil blend on your pet's collar to drive away fleas and other parasitic insects.

Ingredients:

1 cup distilled/purified water

5 drops Lavender Oil

2 drops Citronella Oil

2 drops Purification Oil

2 drops Thieves Oil

1 drop Cedarwood Oil

Combine all ingredients in a bowl. Soak your dog's cotton bandanna into the oil blend and let it dry before using. You may also add a few drops on your dog's collar to repel lice.

35 — Tick Removal

Any pet parent who has experience with ticks will know they are hard to pull off your dog. Use this oil blend to weaken ticks so you can remove them easily.

Ingredients:

2-3 drops Palo Santo

1 drop Rosemary Oil

Apply Palo Santo on the tick for easy removal. It will die and fall off typically within 15-20 minutes. Then, apply Rosemary Oil on your dog's skin, especially on the area where the tick had been. Rosemary acts as an antibacterial and helps keep parasites off your pet.

36 — Tick Repellant

If ticks are a big problem where you live, then this natural tick repellant is what you need. Use it just like a flea repellant by spraying it onto your dog's body.

Ingredients:

120ml base oil (Jojoba Oil, Sweet Almond Oil, Olive Oil, etc.)

6 drops Lemon-Eucalyptus Oil

8 drops Geranium Oil

10 drops Lavender Oil

Mix the oils and apply it to your dog's neck, chest, legs, back and tail.

37 — General Insect Repellant

If you love spending time outdoors with your dog then you'll know how spring and summer can wreak havoc for your furry pal. Insect of all kinds come out and your curious dog can't stop getting bitten by them. Use this oil blend to keep insects off your dog.

Ingredients:

2 cups distilled/purified water

8 drops Peppermint Oil

8 drops Lavender Oil

Mix water and oils in a glass spray bottle. Spritz your dog every day with the mixture, avoiding nose and eyes. You may also spray his bedding or clothes to repel insects even while he sleeps.

38 — Mosquito Repellant

Use this recipe to keep the mosquitoes away from your dog and prevent diseases caused by such insects. Dogs with short fur are the most prone to mosquito bites as they have no difficulty biting skin.

Ingredients:

8 oz. Aloe Vera juice

5 drops Rose Geranium

5 drops Lemongrass Oil

7 drops Citronella Oil

10 drops Myrrh Oil

Mix all ingredients in a spray bottle and spray it on your dog's body before going to a mosquito prone location(avoid contact with eyes). You can also apply it on beddings and around the doorway to repel insects.

39 - Relieve Bee Stings

Bee stings are a painful and annoying, but usually not dangerous event for dogs. If you can remove the stinger safely, do so, and apply a cold compress for a few minutes, then this salve. Apply the ice afterward too to keep swelling down. Swelling is a natural response, but it impedes the healing process. Chamomile and baking soda will work together as anti-inflammatories as well.

Ingredients:

30 mls Coconut Oil

1 tsp Baking Soda

3 drops Vetiver Oil

3 drops Spearmint Oil

1 drop Chamomile Oil

Mix ingredients well and dab onto the sting.

RECIPES FOR CALMING & RELAXATION

40 — Calming Mist #1

A soothing oil blend to help your dog relax. Great for situations your dog tends to feel nervous and anxious about, such as car rides, vet visits and new situations.

Ingredients:

1 cup distilled/purified water

5 drops Lavender Oil

5 drops Roman Chamomile

5 drops Rosemary Oil

Combine all ingredients in a spray bottle and spray on your dog's coat (avoid spraying on his face). You may also rub a few drops between your palms and massage it on your dog's neck, chest and back.

41 — Calming Mist #2

Some dogs struggle to stay calm in new environments or situations. This blend will help your dog relax and stay calm, by helping the areas of the brain that is overly stimulated to become less active.

Ingredients:

1 cup distilled/purified water

5 drops Lime Oil

5 drops Lavender Oil

Mix all ingredients in a spray bottle. Spray it around your dog and avoid spraying directly on his face. You can also spray it on your palms and massage your dog's neck, chest and back.

42 — Calming Oil Blend

This is a simple essential oil recipe to help with anxiety in dogs. If your dog seems to get anxious in certain situations that causes behavioral problems, such as fearful submission (hiding in a corner, body low to the ground, unable to move etc), or fearful aggression (growling, snarling or barking etc), then this blend will help your dog to overcome this emotional barrier.

Ingredients:

8 tbsp carrier oil (Fractionated Coconut oil or Sweet Almond Oil)

2 drops Ylang Ylang Oil

2 drops Clary Sage

2 drops German Chamomile

Mix the oils together. Apply 1/4 teaspoon of the blend to your palms and gently massage into your dog's body. Administer essential oils at least 4 hours before an event that will usually make your dog feel anxious (e.g. environmental changes).

43 — General Fear & Anxiety

If you find your dog doesn't favor the strong smell of Clary Sage, you can try this recipe to calm your pet down whenever he's anxious or become fearful of things.

Ingredients:

2 oz. Jojoba Oil

4-6 drops Lavender Oil

6-8 drops Petitgrain Oil

8-10 drops Neroli Oil

Rub 2-3 drops of the oil blend on your palms and massage it on your dog's toes, ears, armpits and inner thighs.

44 — Recipe for Anxiety

Use this recipe to calm your dog who has noise/separation anxiety or fear of new people, things or places. Separation anxiety is common for dogs that have gotten use to having people around them, and think this is the normal state life should be. If you find your pup having trouble adjusting to being alone, then try using this blend to help ease this adjustment period.

Ingredients:

120ml base oil (Jojoba Oil, Sweet Almond Oil, Olive Oil, etc.)

4 drops Sweet Marjoram

4 drops Clary Sage

8 drops Lavender Oil

8 drops Valerian Oil

Rub 2-3 drops of the mixture between your palms and massage it between your dog's toes, on the edge of his ears and on his armpits and inner thighs.

45 — Recipe for Hyperactivity

All dogs need to be exercised daily and their need will greatly vary from one dog to the next. However, some dogs can get hyperactive even after sufficient exercise when they are overly stimulated. Use this aromatherapy recipe to calm your hyperactive dog.

Ingredients:

120ml base oil (Jojoba Oil, Sweet Almond Oil, Olive Oil, etc.)

3 drops Bergamot Oil

4 drops Sweet Marjoram

5 drops Roman Chamomile

6 drops Lavender Oil

6 drops Valerian Oil

Rub 2-3 drops of the oil blend between your palms and apply it on your dog's inner thighs, on his armpits, between his toes and on his ear tips.

46 — Recipe for Separation Anxiety

Try this simpler oil blend if your pet experiences separation anxiety when you leave him alone at home. This blend doesn't use a base oil to dilute the essential oils so your dog may not appreciate excessive use of this blend if you plan to use this regularly.

Ingredients:

8-10 drops Sweet Orange Oil

4-6 drops Lavender Oil

4-6 drops Ylang Ylang Oil

Combine the oils and rub 2-3 drops between your palms. Massage on your dog's armpits, inner thighs, ear tips and between his toes.

47 — For Relaxation

This is a quick and easy blend to calm your anxious or over-excited dogs. The main active ingredient here is the lavender oil. This blend is more suitable for when you're at home as this is not a topical blend, or if your dog will be in an indoor area.

Ingredients:

1 cup distilled/purified water

3 drops Lavender Oil

1-2 drops Lime Oil

Mix the oils together and pour it in a diffuser. Turn the diffuser on and you will find your pets relaxing and most possibly sleeping within 30 minutes.

48 — Calming Lavender Powder

Make a calming lavender powder for your dog who has stress or anxiety issues. Incorporate baking soda, cornstarch or rice flour to this mixture to create powder. It also works great as a quick dry bath if you use baking soda for this mix.

Ingredients:

1 part Ylang Ylang Oil

2 parts Clary Sage

2 parts Bergamot Oil

3 parts Lavender Oil

Add 12-15 drops of this oil blend to every cup of baking soda and mix well. When your dog is stressed, sprinkle powder on his blanket to help him calm down.

49 - Acute nervousness

At times of extreme stress, it can be very helpful to have a special calming scent that you and your dog can always return to. The sense of smell is deeply linked to memory, not only in humans, but many experts suspect this is true for dogs as well! Calming Valerian and Rose Geranium combine with Sesame Oil, which normalizes the nervous system response, to create soothing oil.

Ingredients:

30 mls carrier oil, such as Sweet Almond Oil

2 drops Valerian Oil

2 drops Rose Geranium Oil

2 drops Sesame Oil

Try dabbing this oil on yourself during calm, peaceful playtime with your pup, and the scent will remind them that everything's okay when they get stressed.

50 - Low Energy After Illness

Frequently, even after the "corner has been turned" of an illness, the usual vim and vigor a dog usually exhibits will be absent for a few days. This spritz is energizing and disinfectant. This is important because the immune system may still be a little low during this time, and needs to be protected.

Ingredients:

240 mls Water

15 drops ethanol (either vodka or perfumers oil will do)

1-2 drops Ylang Ylang Oil

1-2 drops Lemongrass Oil

1-2 drops Tangerine Oil

Spray this around the room your dog will be spending time in during recovery.

51 - Restlessness

Especially in "teenage" dogs, the ones who are not puppies anymore, but not quite full-grown, we frequently see them have trouble settling down. This balm will settle the nervous system and allow jumpy muscles to relax while soothing the emotions as well. Massage it into the chest and between the forelegs to provide maximum calming and balancing effects.

Ingredients:

30 mls Extra Virgin Coconut Oil

10 mls Argan Oil

1-2 drops Ylang Ylang Oil

1-2 drops Bergamot Oil

1-2 drops Clary Sage Oil

Mix together thoroughly and keep in a small jar. The texture should be like a thick serum, which makes it light enough to massage in quickly.

52 - Sleep Support

If your pet has a consistently hard time getting to sleep, even when they are clearly tired, that might be something a veterinarian should look at. If the sleep troubles are only intermittent and not caused by any underlying issue though, this diffusion can help. Valerian is a time-honored sleep inducer, and St. John's Wort calms the nervous system to help your pet stay asleep.

Ingredients:

1 Cup Water

1 drop Lavender Oil

1 drop Valerian Oil

1 drop Chamomile Oil

1 drop St. John's Wort Oil

Pour into a diffuser and turn diffuser on for 45 minutes before bedtime.

53 - Fear of car rides

A little bit of simple training combines with powerful soothing oils to banish fear of car rides. Dab a little of this oil on your body and hold your dog in calm, peaceful playtimes for a week or so. Then, when it's time to get in the car, massage it into their chest and neck. The association of happy times will calm them, while Valerian works on the nervous system to settle it.

Ingredients:

30 mls Carrier Oil, such as Jojoba or Coconut Oil

4 drops Valerian Oil

2 drops Ginger Oil

Mix thoroughly and store in a small jar at room temperature.

54 - Crate Anxiety

This is another example of how training and oils can work together. Spray this mixture on a favorite blanket and incorporate it into gentle playtime. Then, an hour before it's time to get into the crate, spray a small spritz of this inside. The association will work together with Lavender to calm the nerves. Clary Sage has been shown to boost mood and increase happiness and "home" feelings in humans. Some experts think this may work for canines as well.

Ingredients:

120 mls Water

2 drops Lavender Oil

1 drop Clary Sage Oil

Mix in a spray bottle and store in the refrigerator. Shake well before using.

RECIPES FOR JOINT & MUSCLE HEALTH

55 — Sore Joint Oil Blend #1

This recipe will soothe your pet's sore joints and ease pain caused by dysplasia or arthritis. Ginger and lemon is great for inflammation of tissue and joints while lavender can help calm the area down.

Ingredients:

120ml base oil (Jojoba Oil, Sweet Almond Oil, Olive Oil, etc.)

8 drops Ginger Oil

6 drops Lavender Oil

8 drops Lemon Oil

Mix all the ingredients. Apply the mixture topically on your dog's swelling joints and massage gently. This is best used daily on the required area.

56 — Sore Joint Oil Blend #2

Another oil blend that will ease pain and swelling in joints.

Ingredients:

120ml base oil (Jojoba Oil, Sweet Almond Oil, Olive Oil, etc.)

4 drops Peppermint Oil

7 drops Valerian Oil

5 drops Ginger Oil

8 drops Helichrysum Oil

Combine all ingredients and massage the mixture on the affected areas. You may also apply 1-2 drops on the inside of your dog's ear tips for added comfort.

57 — Arthritis Relief

This oil blend provides great comfort for pets with joint pains or arthritis. Birch oil is great for muscle aches and pains, but also very effective for chronic pain such as arthritis.

Ingredients:

7 drops Rosemary Oil

8 drops Juniper Oil

12 drops Birch Oil

Mix the oils together and apply directly on painful and sore joints. Massage the affected area twice per day to soothe the pain away.

58 — Arthritis Oil Blend

This blend uses frankincense oil to actively treat arthritis. Frankincense has long been used as a cure for arthritis in India and has recently been scientifically proven for it's healing properties. However do consult your vet before use as the amount of frankincense oil used should correlate to the weight of your dog.

Ingredients:

½ oz. Fractionated Coconut Oil

6 drops Frankincense or Helichrysum Oil

4 drops Peppermint Oil

3 drops Vetiver Oil

2 drops Ginger Oil

Mix the oils and use topically to relieve sore joints and pain caused by arthritis. Rub a few drops on your palms and massage into the affected areas. You can also apply a few drops on your dog's ear tips for extra comfort.

59 — Rheumatism Oil Blend

As they age, dogs suffer from rheumatism just like humans. Prepare this blend to help you aging pal cope with the swelling of joints.

Ingredients:

7 drops Rosemary Oil

8 drops Birch Oil

8 drops Juniper Oil

Pour these oils in a 10-ml dark glass bottle and mix well. Apply 2-4 drops to your hands and massage your dog in the neck, chest and back to alleviate pain. Do this in the morning and at night before sleep.

60 — Muscle Ache Ointment

This mixture helps soothe muscle aches and pain in dogs. Some breeds that are prone to hip dysplasia will find this blend helpful in soothing the thigh muscles.

Ingredients:

1 tbsp Fractionated Coconut Oil

3 drops Lavender Oil

2-3 drops Copaiba Oil

Combine all the oils and store in a glass bottle. Whenever your dog is showing symptoms of sore muscles, rub the mixture on the affected area(s) for immediate relief.

61 — Dog Paw Ointment

Use this recipe to soothe and soften your dog's paws. Extreme weather conditions can cause damage to your dog's paw pads, causing them great discomfort. This balm will help protect and sooth in hot, dry and cold weather.

Ingredients:

2 tbsp Fractionated Coconut Oil

2 tbsp Shea butter

2 tbsp beeswax

1 teaspoon Jojoba Oil

1-2 drops Lavender Oil

1-2 drops Thieves Oil

1-2 drops Frankincense Oil

Combine Coconut Oil, Shea butter and beeswax in a small glass. Sit the glass in a pot with some water and place over low heat. Once melted and cooled, add the essential oils then transfer in a container.

62 - **Muscle Soreness**

This oil blend is particularly effective for sore muscles caused by large amounts of exercise or exertion. Even young, healthy dogs can be left feeling stiff after a long hike or swim, and this will help. Because the oils used are quite strong and invigorating, use this only for a day or two at a time and then stop. This should not be used on animals that are known to have issues with dry or sensitive skin.

Ingredients:

30 mls Fractionated Coconut Oil

1 drop Juniper Oil

1 drop Cyprus Oil

1 drop Eucalyptus Oil

1 drop Orange Oil

1 drop Peppermint Oil

Blend oils together well and rub vigorously into sore muscles. One application should be sufficient to soothe sore muscles.

63 - Painful Joints

Joint pain isn't always caused by arthritis or rheumatism. Sometimes during injury recovery or just from exertion, temporary joint pain is the result. This oil will soothe the pain and also encourage recovery.

Ingredients:

30 mls Carrier Oil, such as Jojoba or Coconut

2 drops Turmeric Oil

2 drops Vetiver Oil

2 drops Ginger Oil

Massage lightly into affected joints daily, for up to a week.

64 - Sore Spine

Back pain is particularly hard to bear, even for people. Our achy dogs might not be able to tell us when they have a sore back, but it's our job to know! Watch for a usually jumpy dog that seems reluctant to move around, and massage this oil directly into the skin above the spine. As always, make sure that if the symptoms persist more than a few days you get a professional opinion.

Ingredients:

30 mls Carrier Oil, such as Jojoba or Coconut Oil

1-2 drops Wintergreen Oil

1-2 drops Cypress Oil

Mix all ingredients well and store in a small jar.

65 - Ligament Pain

Both injury and age can cause the particularly gnawing pain that happens in the ligaments. The Black Pepper Oil in this blend increases circulation, while Lemongrass and Sweet Almond Oil are soothing. A few minutes of massaging this into the affected joint once a day for up to a week should ease the pain as the joint heals itself.

Ingredients:

30 mls Carrier Oil, such as Jojoba or Coconut Oil

2 drops Lemongrass Oil

2 drops Black Pepper Oil

10 drops Sweet Almond Oil

Mix all ingredients thoroughly and store in a small jar

RECIPES FOR EAR, NASAL & DIGESTIVE HEALTH

66 — Ear Infection Ointment #1

This oil recipe will help prevent and treat common ear infections among dogs. If your dog is prone to earwax building up in their ear canal then you can use this after cleaning their ears once a month.

Ingredients:

120ml base oil (Jojoba Oil, Sweet Almond Oil, Olive Oil, etc.)

5 drops Bergamot Oil

5 drops Niaouli Oil

6 drops Roman Chamomile

8 drops Lavender Oil

Combine the oils in a glass bottle and mix well. Massage the outside of your dog's ear with the oil blend and apply a few drops into his ear canal using a dropper.

67 — Ear Infection Ointment #2

Another oil recipe that will help treat ear infection and prevent it from coming back.

Ingredients:

2 tbsp Fractionated Coconut Oil

10 drops Arborvitae

15 drops Basil Oil

15 drops Frankincense Oil

15 drops Geranium Oil

15 drops Lavender Oil

Combine all ingredients in a glass bottle and shake well to mix. Use this oil blend once per week to treat ear infection or once per month to prevent recurrence of the infection.

68 — For Ear Health

This oil blend encourages healthy function in your dog's ears. Lavender oil is gentle and effective in cleaning out dirt from your dog's ears.

Ingredients:

1 tbsp Fractionated Coconut Oil

5 drops Geranium Oil

5 drops Melaleuca Oil

5 drops Lavender Oil

Mix all ingredients and store in a glass bottle. After cleaning your dog's ears with an all-natural cleaner, use a cotton bud to rub a few drops in his ears. Don't push the bud in where you cannot see it. Do this twice per day until you see improvement.

69 — For Sinus Infections

This oil blend is effective in relieving nasal congestion and allows your dog to breathe more easily. This is great for brachycephalic dog breeds who can experience breathing difficulties in hot weather.

Ingredients:

120ml Sweet Almond Oil

2 drops Niaouli Oil

4 drops Myrrh Oil

8 drops Eucalyptus Oil

Combine all ingredients in a glass bottle. You can either massage several drops into the neck and chest of your dog or apply it to his bandanna. You can also add several oil drops on his bedding to relieve nasal congestion.

70 — Tummy Ache Ointment

Use this recipe to help ease upset stomach and bring back your dog's vivacity. DiGize oil is famous for it's effectiveness to regulate the digestive system and sooth digestive pain.

Ingredients:

1 tbsp Fractionated Coconut Oil

3 drops Peppermint Oil

2-3 drops DiGize Oil

Combine the oils in a glass bottle. If you see your dog exhibiting signs of an upset tummy, rub a few drops on your dog's belly and massage gently with your hands.

71 — Flatulence Oil Blend

It may be comical to hear your dog toot, but just like humans, dogs can have painful uncomfortable gas. This oil blend can greatly help in alleviating flatulence caused by excess gas.

Ingredients:

15ml base oil (Sweet Almond Oil, Jojoba Oil, Hazelnut, etc.)

3 drops Tangerine Oil

3 drops Nutmeg Oil

3 drops Cinnamon Oil

3 drops Cardamom Seed Oil

3 drops Caraway Oil

Combine the oils and store in a dark glass bottle. Place 2 drops on your dog's food and administrate another 1-2 drops after eating.

72 — For Motion Sickness

Use this oil blend to calm the stomach of your dog if they are prone motion sickness.

Ingredients:

120ml base oil (Jojoba Oil, Sweet Almond Oil, Olive Oil, etc.)

10 drops Peppermint Oil

14 drops Ginger Oil

Apply the mixture to your dog's ear tips, belly and armpits. When travelling, add a few oil drops to a cotton ball and place it in front of your car's air vent to allow the scent to circulate.

73 - Heavy Breathing

Sometimes for what seems like no reason, our dogs' breathing becomes slightly labored and noisy. Usually it is just a little sinus trouble that can be soothed by inhaling this blend of anti-inflammatory sandalwood, eucalyptus, and peppermint diffused into the air for 30 minutes. If your dog's breathing issues persist though, make sure to take them in to the vet, as this can sometimes be a sign of bigger problems.

Ingredients:

4 drops Sandlewood Oil

2 drops Frankincense Oil

2 drops Eucalyptus Oil

1 drop Peppermint Oil

1 cup Water

Pour into your diffuser and turn on the diffuser for about 30 minutes.

74 - Digestion Trouble

Trouble digesting food can come from many sources. We frequently see this in animals who are recovering from more severe stomach ailments. Oils can encourage proper digestion simply through inhalation. When the food is being properly digested, appetite should increase and demeanor should also improve. Remember, if your dog appears to have trouble digesting for more than a few days, take them into the vet for examination.

Ingredients:

1-2 drops Ginger Oil

1-2 drops Fennel Oil

Apply these oils to a bandana and tie around your dog's neck with the oiled side facing outward, so the scents are easily inhaled and no direct skin contact is being made.

75 - Ear ache Massage Oil

This oil is not designed to treat ear infection, as some of the other recipes are. This can be used in conjunction with those just to ease pain and inflammation, and facilitate massage that can ease the pressure and discomfort that goes along with earache. Even without an infection, ear pain is not uncommon. If you see your dog rubbing its ears on the floor or other surfaces, this can help soothe that discomfort.

Ingredients:

30 mls Carrier Oil, such as Coconut Oil or Jojoba

1-2 drops Cassia Oil

1-2 drops Frankincense Oil

Mix ingredients thoroughly and massage directly on the ears.

76 - Heartburn

Yes, dogs get heartburn too! Some of the signs that your pet has heartburn or acid reflux are spitting up food, whining/howling while swallowing, and hacking. If this persists for more than a few days, the dog should be seen by a veterinarian. For a day or two at a time though, this might just do the trick. Since this is taken orally, be sure to keep the amount of oil in proportion to your dog's body size. Dogs under 30 lbs should use the smaller amount.

Ingredients:

1-2 drops Chamomile Oil

1-2 drops Marjoram Oil

Add to no more than one meal per day.

77 - Seasonal allergies

The uncomfortable stuffiness that accompanies hay fever is just as uncomfortable for our four-legged friends as it is for us. If your pet doesn't tolerate a diffuser well, this salve can be rubbed into the fur of the chest and between the forelegs, and the oils will waft through the nose and sinuses.

Ingredients:

30 mls Carrier Oil, such as Jojoba or Coconut Oil

1 drop Peppermint Oil

1 drop Juniper Oil

1 drop Rosewood Oil

Blend thoroughly and massage into your dog's chest to relieve stuffed-up sinuses.

78 - Sinus Inflammation

Signs of sinus inflammation might be things like snoring or loud breathing during the day. This could be caused by something more serious, and if it persists past a week you should take your dog into the vet to make sure everything is ok. If it's something simple like dryness in the air or sensitivity to something in their surroundings though, this diffusion will work wonders.

Ingredients:

1 cup Water

1 drop Eucalyptus Oil

1 drop Peppermint Oil

1 drop Turmeric Oil

1 drop Frankincense Oil

Pour ingredients into your diffuser and turn it on for 30 minutes. The smell can be quite strong at first, so make sure your dog has a way to leave the room if it becomes overwhelmed.

79 - Decrease Appetite

As sad as it is, sometimes our furry friends gain a little more weight than is healthy for them. If your veterinarian has prescribed a weight-loss strategy for your little buddy, this massage oil can make the process a little easier and less frustrating for you both. Massage this into the skin and fur of the chest and between the forelegs between meals to stave off appetite.

Ingredients:

30 mls Carrier Oil, such as Jojoba or Coconut Oil

2 drops Grapefruit Oil

1 drop Peppermint OIl

1 drop Fennel Oil

Mix all ingredients well and store in a small jar.

80 - Doggy Breath

As much as we love them, we can all admit that our pups' breath doesn't always smell the freshest. Persistent halitosis can be a sign of underlying problems, so if the problem lasts more than a week or so, make sure to take them in. If not, this mixture added to a treat or biscuit can help greatly. Because this is taken internally, make sure to monitor the amount very carefully.

Ingredients:

1 drop Peppermint Oil

1 drop Cassia Oil

Place mixture on a treat or biscuit once a day, for no more than 4 or 5 days.

81 - Vomiting

Obviously vomiting can be caused by some really unpleasant things, so make sure your dog gets the veterinary attention they need if they are vomiting. If the vomiting and nausea is something that just has to be waited out, this remedy can help. Just dab these oils on a bandana and tie around the dog's neck. The olfactory effects they have reduce nausea.

Ingredients:

1 drop Ginger

1 drop Cardamom

1 drop Peppermint

Tie around the neck with the oil facing out toward the room, away from the skin.

RECIPES TO BOOST ENERGY & IMMUNE SYSTEM

82 — For Energy Boost

This oil blend will boost your dog's energy and keep him energized all throughout the day.

Ingredients:

2 drops Peppermint Oil

5 drops Rosemary Oil

6 drops Lavender Oil

Pour the essential oils in a glass bottle and mix well. Massage a few drops on your dog's spine every morning to give him a good head start.

83 — Fatigue Oil Blend

Use this oil blend to revitalize your dog who is suffering from malaise and fatigue caused by an illness, old age or overactivity.

Ingredients:

15ml base oil (Sweet Almond Oil, Jojoba Oil, Hazelnut, etc.)

3 drops Ylang Ylang Oil

6 drops Tangerine Oil

7 drops Rosemary Oil

Combine oils in a glass bottle and mix well. Add 2-4 drops to your dog's food as needed. Be cautious in using Rosemary if your dog is prone to seizures.

84 — Increasing Appetite

If you notice your dog is losing appetite due to either sickness or old age, give him this oil blend to make him excited about food again.

Ingredients:

15ml Sweet Almond Oil

2 drops Lemon Oil

2 drops Sweet Orange Oil

2 drops Bergamot Oil

2 drops Grapefruit Oil

2 drops Lime Oil

Mix all ingredients together in a dark glass bottle. Pour 2-6 drops into your hands and massage your dog in the neck and chest.

85 — Immune System Oil Blend

This recipe may help improve your dog's immune system as the oils naturally have medicinal properties.

Ingredients:

1 cup distilled/purified water

5 drops Lavender Oil

5 drops Frankincense Oil

5 drops Roman Chamomile

Mix all ingredients in a bottle and spray it on your dog's body but not on his face. You may also apply a few drops of the oils on your palms and massage your dog's chest, back and neck area.

86 — Immune Support

Besides giving them a grain-free diet, add this simple oil blend to your dog's food to encourage strong immune system.

Ingredients:

1 drop Lavender Oil

1 drop Lemon Oil

1 drop Peppermint Oil

Combine the oils and mix well. Place it in a capsule to give to your dog or put it in his food for a healthier repast.

87 - Fever

Dogs' internal body temperatures naturally run higher than in people, so don't assume your dog is feverish just because they feel a little warm. If, however, your pet has been diagnosed with a fever due to some infection or other condition by a vet, this ointment can help to alleviate it. Make sure the underlying cause of the fever is being treated; this remedy is just to provide comfort and boost immunities while your dog recovers.

Ingredients:

30 mls Carrier Oil such as Jojoba or Coconut Oil

3 drops Garlic Oil

2 drops Eucalyptus Oil

1 drop Peppermint Oil

Mix ingredients well and massage into the neck and chest.

88 - Headache

Some symptoms of headache in dogs are low energy, avoiding sounds more than usual, and rubbing the head or ears on walls or other surfaces. If it's just a sporadic thing it's usually nothing to worry about, but if this happens frequently, make sure to take your dog in to make sure there no serious underlying cause. This ointment can help relieve pressure and pain while providing immune protection to prevent your pet from getting sick while its immune system may be low.

Ingredients:

30 mls Carrier Oil such as Jojoba or Coconut Oil

3 drops Spearmint Oil

2 drops Helichrysum Oil

2 drop Lavender Oil

1 drop Roman Chamomile Oil

Mix ingredients well and massage into the neck and spine.

89 - Antibacterial Spray

This is a great all-purpose spray to have in your refrigerator. It can be used as an air spray, on dog beds, or on the coat if there is concern about bacterial exposure. This is a nice first line of defense against getting sick if you've had a sick visitor, or been around other dogs who are sick. Always check sprays meant for fabrics in an inconspicuous spot before use. This also boosts the immune system to prevent infection.

Ingredients:

120 mls Witch Hazel

15 mls Baking Soda

1-2 drops Lemongrass Oil

1-2 drops Palmarosa Oil

Combine in a spray bottle and shake before using.

RECIPES FOR NAIL & FUR HEALTH

90 - Nail Health

Your dog's nails aren't just cute little fingernails for pups! They are also important tools dogs use to dig and explore the world. Because they are exposed to so many irritating and drying conditions down there on the ground, it's nice to keep them healthy with a cuticle massage after bath time. This serum is perfect for keeping a healthy pup's nails as healthy as it is.

Ingredients:

10 ml Carrier Oil, such as Sweet Almond or Jojoba

1 drop Carrot Oil

1 drop Balsam Fir Oil

Mix well in a dropper bottle, and drop one drop on each nail, massaging it in as you go.

91 - Nail Antifungal

Signs of fungus at the root of the nail can be peeling skin, flaky cuticles or a dusty appearance to the nail (even when the nails are clean). If this condition persists for more than a week or two, your dog should be seen by a vet, but this serum might be what it takes to sort it out. Lemon and Geranium oil are wonderful antifungal agents, and Coconut oil will replace moisture lost to the infection.

Ingredients:

10 mls Fractionated Coconut Oil

1 drop Lemon Oil

1 drop Geranium OIl

Mix ingredients together in a dropper bottle. Apply a drop to each nail, massaging as you go.

92 - Weak nails

There are many possible underlying causes for the nails becoming weak. Some of them can be quite serious, so if your dog suddenly develops this issue you should have them checked out. On the other hand, some dogs naturally don't have the sturdiest of nails. This serum will help strengthen them.

Ingredients:

30 mls Sweet Almond Oil

2 drops Geranium Oil

2 drops Lemon Oil

5 drops Jojoba OIl

Mix together in a dropper bottle. Apply one drop to each nail after bath time, massaging as you go.

93 - Hair Loss

Whether your dog has developed bald spots because of skin issues or injury or he's had to be shaved because of surgery, you can encourage the hair to regrow quickly and healthier than ever. Note: Make sure the underlying cause of the hair loss has been addressed. Depending on what it is, it could seriously affect the health of your pet.

Ingredients:

30 mls Grapeseed Oil

10 mls Jojoba Oil

2-3 drops Cedarwood Oil

2 drops Rosemary Oil

2 drops Lavender Oil

Mix all ingredients and massage into affected area a few times a week for up to two weeks.

94 - Hair Thinning

This differs from localized hair loss, because it's an all-over thinning of the coat. Again, make sure there is no serious underlying cause, and then you're free to use the all-natural and highly effective spray. The Cedar and Sage oil work together on the skin to encourage regrowth, and the Rosemary Oil ensures that the hair grows back healthy and strong.

Ingredients:

240 mls Water

5 drops Cedar Oil

5 drops Sage Oil

5 drops Rosemary Oil

Mix all ingredients in a spray bottle and store in the refrigerator. Spray all over the coat, avoiding the face and head, two or three times a week for up to two weeks.

95 - Seborrhea (oily skin)

Just like people, dogs can have oily, greasy skin. It isn't usually a health concern, but it can cause the fur to become lank and greasy, and oil trapped on the skin's surface can cause irritation and even pustules. This spray will help. Spray it all over the skin, parting the hair if necessary, and massage it in thoroughly. Most animals respond wonderfully to treatment once a week.

Ingredients:

100 mls Witch Hazel

2 drops Neroli Oil

2 drops Argan Oil

1 drop Lemon Oil

Mix in a spray bottle and store in the refrigerator. This mixture will keep for about two weeks.

96 - Burr Repellant

After a fun day out in the woods or fields, your dog may be happy, but he will almost definitely be covered in burrs. This oil is a great trick for preventing the long minutes of painful grooming to get them out. Simply massage this oil throughout the fur, especially on the legs and sides before a day out in the wilderness. Fewer burrs will stick, and the ones that do will come out much more easily.

Ingredients:

40 mls Fractionated Coconut Oil

4 drops Almond Oil

4 drops Sesame Oil

Combine all ingredients and massage into fur.

97 - Excessive Shedding Shampoo

Most animals shed, at least a little bit. This shampoo can reduce that amount, which is a blessing if your pet happens to shed more than usual. Swap this shampoo in for the normal one that you use every third or fourth bath. The essential oils will strengthen the hair roots and save you some grooming time.

Ingredients:

100ml distilled/purified water

100ml Jojoba oil

2 tbsp Coconut Oil

2 tbsp Castile Soap

2 drops Roman Chamomile

2 drops Neem Oil

1 drop Cedarwood Oil

Mix all ingredients thoroughly and shampoo as you normally would.

98 - Fur Detangler

Whether you have a long-haired beauty you brush frequently, or have recently adopted a tangled, messy little buddy, there are times when knots need to be worked out of the fur. This mixture of slippery oils and deodorizing Apple Cider Vinegar works wonders. Oils slide the knots apart and ACV makes sure that any trapped odors disappear right away.

Ingredients:

120 mls Water

40 mls Apple Cider Vinegar

10 drops Jojoba Oil

2 drops Rosemary Oil

1 drop Rose Hip Oil

Mix well in a spray bottle and spray on sections of hair at a time as you brush.

99 – Cracked nails

Dogs love to run and when they run on hard surfaces for a long time, your doggy pal is bound to crack a nail eventually. Damaged nails can cause discomfort or even pain in the worse situation. This recipe can be applied once everyday to help strengthen your dog's nails overtime. You'll notice them not cracking or splitting as often.

Ingredients:

10ml Grapeseed Oil

10ml Jojoba Oil

3 drops of Neem Oil

Mix the ingredients together and store in a cool dry place, out of direct sunlight. You can also add this mixture to melted beeswax to produce a waxing balm once harden.

100 – Shiny Coat

Some dogs are prone to dry flaky skin, which not only makes their coat look dull and brittle, but also causes your pal to be a itchy miserable mess. Use this recipe when you have time to really pour your attention onto your dog. The key ingredient here is oatmeal and Jojoba Oil, where both ingredients help regulate the skin's natural ability to take care of itself. The added lavender oil is added as a calming agent.

Ingredients:

Powdered Oatmeal

30ml Jojoba Oil

3 drops of Lavender Oil

Run a bath with lukewarm water and pour in powdered oatmeal until the water is sufficiently cloudy. You can buy any unsweetened oatmeal and blend it in a blender for a few minutes to get it powdered. Then add the rest of the ingredients in and massage your dog's fur and skin. You'll want your dog to soak in the bathtub for a good 10minutes before towel drying.

CONCLUSION

I hope this book was able to help you resolve all your dog's problems in a safe and effective manner. These 50 simple essential oil recipes can certainly go a long way to maintaining health and wellness of your four-legged companion!

However, these recipes still need to be administrated correctly for your dog to reap the full benefits essential oils can have. Your four legged friend is built very different to you and without the proper knowledge of how essential oils work, you may unintentionally do more harm than good.

I recommend you to fully educate yourself with the basics of essential oils before administrating any of these recipes as dogs come in many shapes and sizes with their own personal health quirks. Educating yourself with more knowledge will do wonders to how well your dog will benefit from these recipes and prevent unwanted harm.

You can purchase my Essential Oils for Dogs guide book to make sure your dog is completely safe during your use of essential oils by equipping yourself with the nessessary knowledge.

Click the link to get it! http://amzn.to/2c5APpN

This book will take you through all the basics of how essential oils work on dogs, the potential they have in various medical and therapeutic applications. You will learn how to properly introduce your dog to essential oils as well as how to safely monitor your dog's use of essential oil. Everything you need to know to get started on safely using essential oils on your dog will be in this one book!

Finally, if you enjoyed this book, then I'd like to ask you for a favor, would you be kind enough to leave a review for this book on Amazon? It'd be greatly appreciated!

ABOUT THE AUTHOR

Julie Summers has dedicated her life to holistic and alternative medicine for our loyal companions. Growing up with a Veterinarian for a mother, animals have always been a large part of Julie's life. She started her journey as far back as she can remember, constantly seeking ways to better care for her animal pals.

Julie started to formally train in aromatherapy and acupressure for animals a decade ago. She received her certification in Natural Health for Animals at the University of Natural Health in 2007.

Outside of being an author, she works as a manager at a dog boarding centre. Employing her deep knowledge of alternative treatments has enabled her to provide outstanding results for all of her clients!

Made in the USA
Lexington, KY
18 October 2016